Numbers 1 to 30

Level K1
Mathematics

Alyse Sweeney

Welcome to studySMART !

SCHOLASTIC

Numbers 1 to 30 reinforces your child's classroom learning of number recognition, number formation, quantities and counting. It provides the practice your child needs to thoroughly learn the numbers 1 to 30.

A variety of engaging activities for numbers 1 through 30 include:
• Tracing and writing numbers
• Identifying and sequencing numbers
• Drawing specified quantities
• Counting objects in a set
• Comparing sets of objects

The systematic and predictable write-and-learn activities encourage independence and feelings of success. Parent notes at the end of the activities provide ideas for extending your child's learning.

Review pages throughout the book and a number chart further support learning.

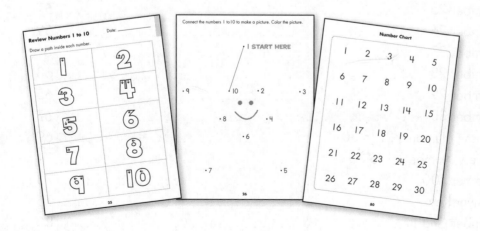

Note: To avoid the awkward 'he or she' construction, the pronouns in this book will refer to the male gender.

Contents

Trace and write.

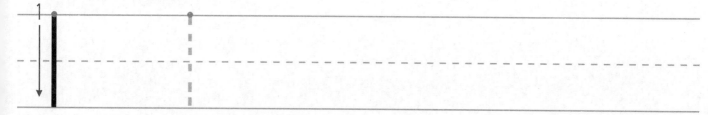

Color each set of 1 crab.

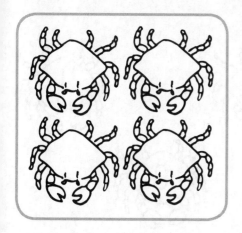 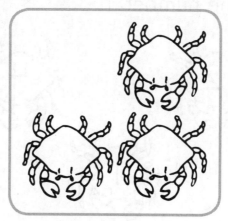 

  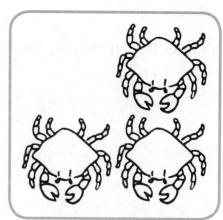

Draw 1 crab.

Color each crab with the number 1.

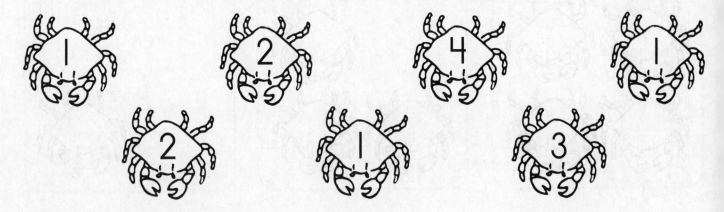

Write the missing number.

_____

_____

2 3

To parents Tell your child that he has **1** nose and 1 mouth on his face. Encourage him to point these out.

6

Date: _____

2   two

Trace and write.

Color each set of 2 lollipops.

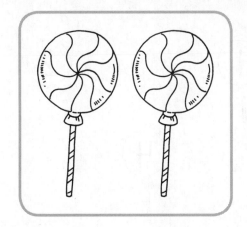 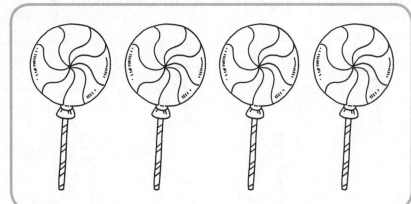

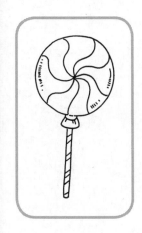 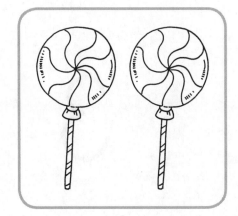 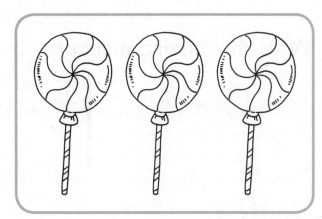

Draw 2 lollipops.

Color each lollipop with the number 2.

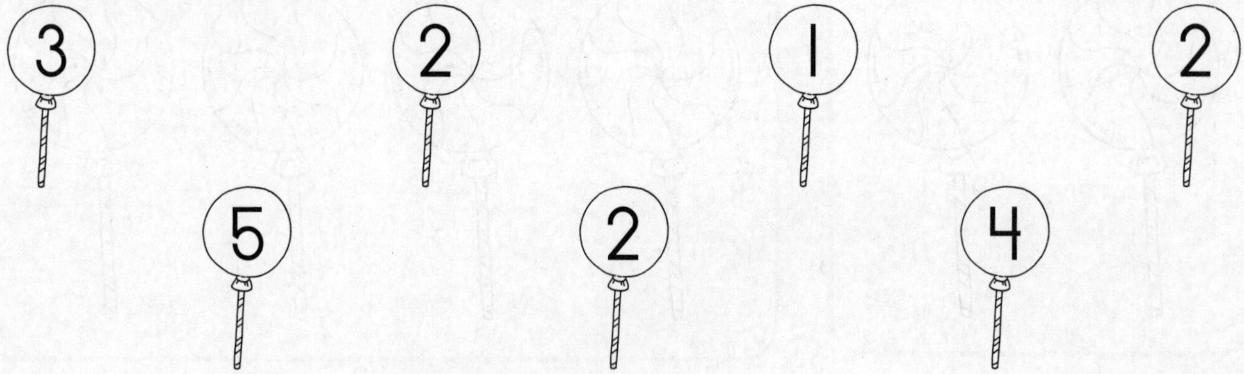

③ ② ① ②

⑤ ② ④

Write the missing number.

1 ____ 3

8

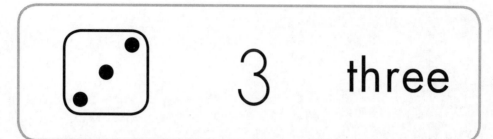

Trace and write.

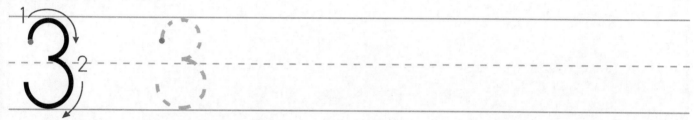

Color each set of 3 flowers.

Draw 3 flowers.

Color each flower with the number 3.

 3   2   3    5

 6    1    3

Write the missing number.

2 _____ 4

Trace and write.

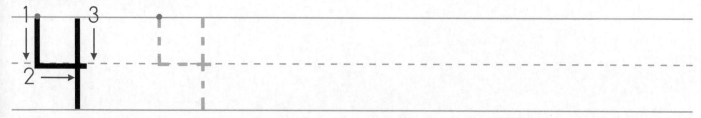

Color each set of 4 bananas.

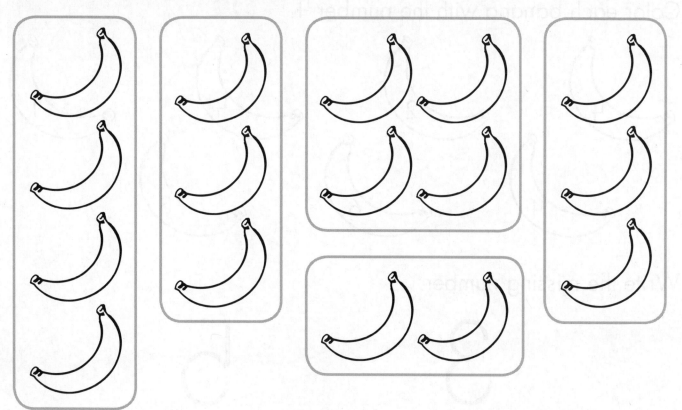

Draw 4 bananas.

Color each banana with the number 4.

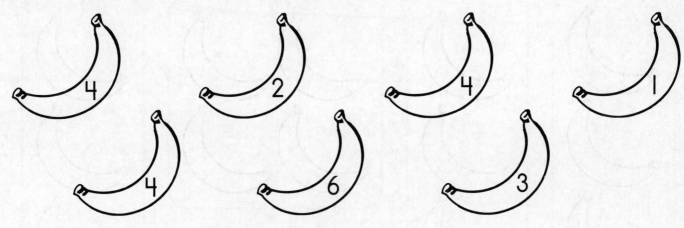

4   2   4   1

4   6   3

Write the missing number.

3 _____ 5

To parents Draw the top of a rectangular table on a sheet of paper. Have your child draw the legs of the table. After he is done, ask him to count the **4** legs.

5   five

Trace and write.

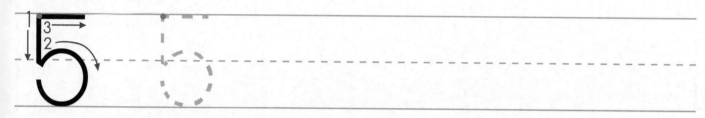

Color each nest with 5 eggs.

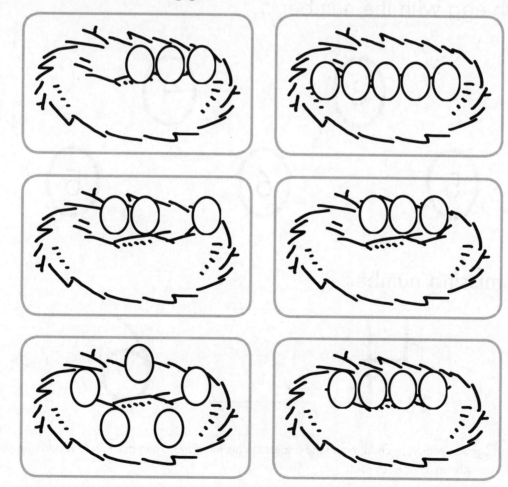

Draw 5 eggs.

Color each egg with the number 5.

5 3 4 2

5 6 5

Write the missing number.

4 ___ 6

To parents Together with your child, count the fingers on his left hand. Then ask him to repeat the process with his right hand or each foot.

6   six

Trace and write.

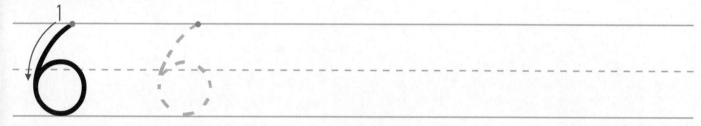

Color each set of 6 balls.

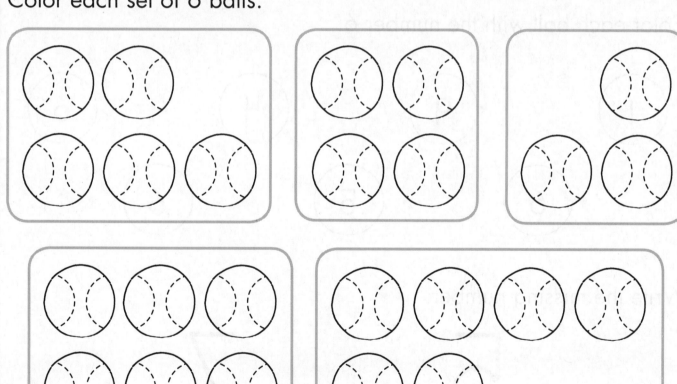

Draw 6 balls.

Color each ball with the number 6.

 ① ⑨ ④ ⑥

⑥ ⑤ ⑥

Write the missing number.

5 _ _ _ _ _ 7

7   seven

Trace and write.

Color each set of 7 planes.

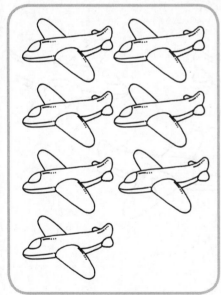

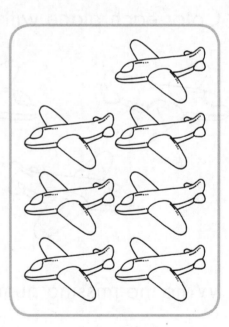

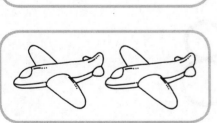

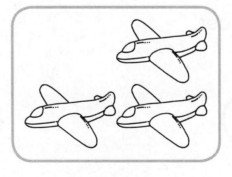

Draw 7 planes.

Color each plane with the number 7.

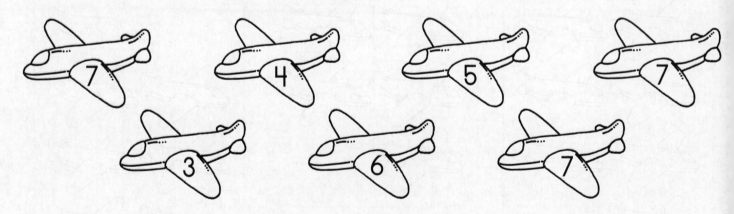

Write the missing number.

6 ------- 8

To parents Go on a 'Number **7** Hunt' around the house with your child. Ask him to find any 7 like objects. For example, 7 apples, 7 pens and so on.

Date: _____

Trace and write.

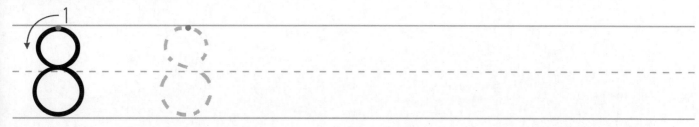

Color the set of 8 houses.

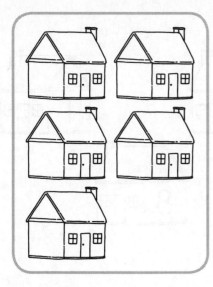

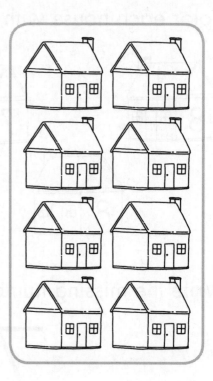

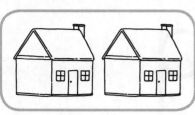

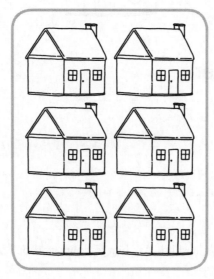

Draw 8 houses.

Color each house with the number 8.

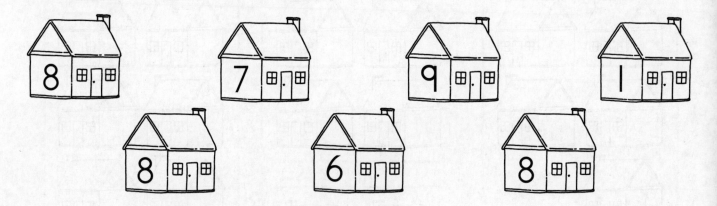

Write the missing number.

7 _ _ _ _ q

To parents   Sit by a bookshelf with your child. Ask him to count and pull out **8** books.

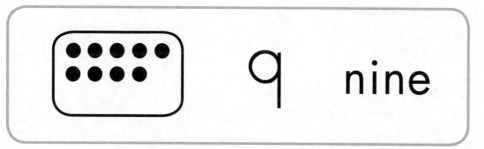

Trace and write.

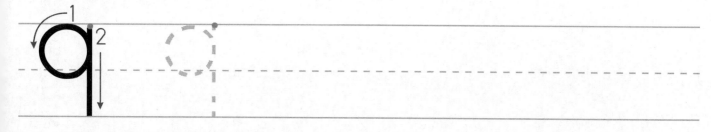

Color the tree with 9 apples.

Draw 9 apples.

Color each apple with the number 9.

Write the missing number.

To parents Take a box of building blocks. Ask your child to count **9** blocks and line them up.

Date: _____

10 ten

Trace and write.

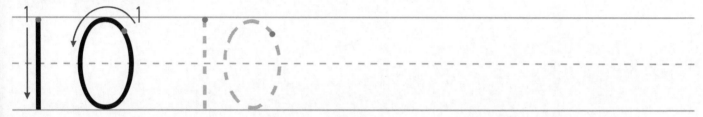

Color the pizza with 10 tomato slices.

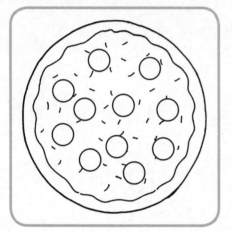

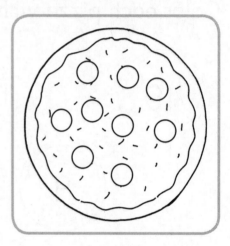

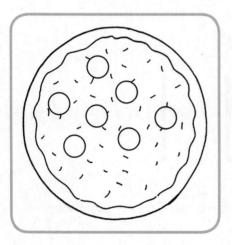

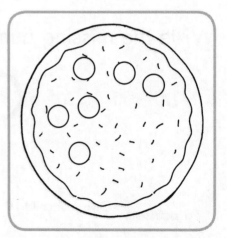

Draw 10 pizzas.

Color each pizza with the number 10.

Write the missing number.

To parents   Give your child **10** cotton balls and ask him to arrange these in a circle. Have him count each cotton ball.

Review Numbers 1 to 10

Draw a path inside each number.

Connect the numbers 1 to 10 to make a picture. Color the picture.

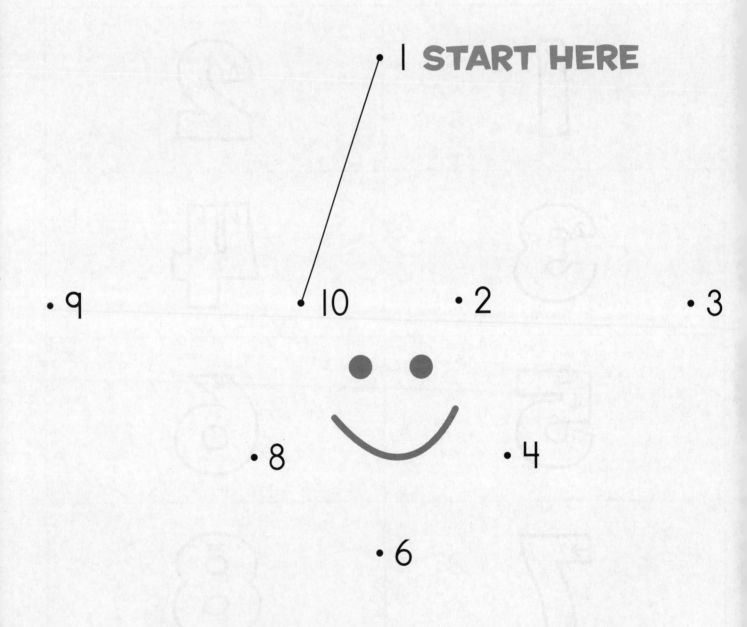

1 START HERE

•9   10•   •2   •3

•8   •4

•6

•7   •5

Trace and write.

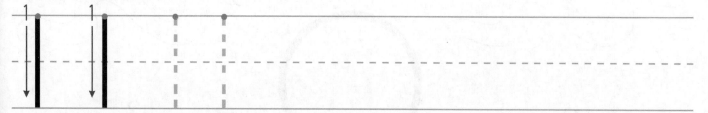

Color the snakes with 11 stripes.

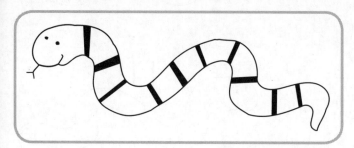

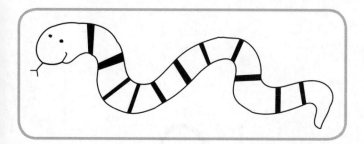
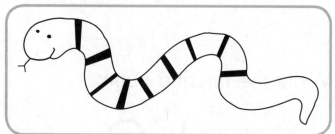

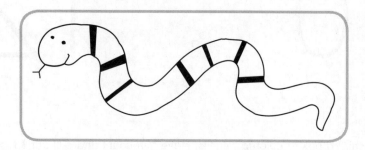

Draw a line from the 11 to each snake with an 11.

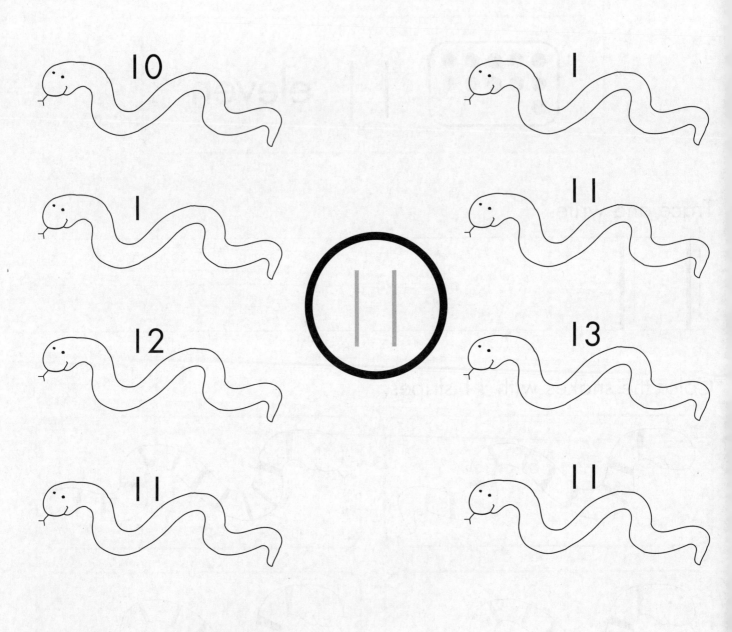

10

1

1

11

12

13

11

11

Write the missing number.

10 ------ 12

Date: _____

 12 twelve

Trace and write.

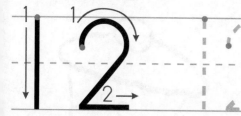

Color the set of 12 cupcakes.

Draw a line from the 12 to each cupcake with a 12.

Write the missing number.

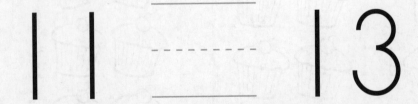

11 ___ 13

To parents Go with your child to the neighborhood fruit-seller. Ask him to pick out **12** apples. Tell him that 12 of something make a dozen.

Trace and write.

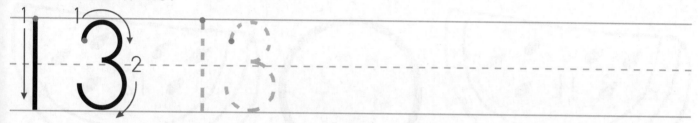

Color the watermelon slice with 13 seeds.

Draw a line from the 13 to each watermelon slice with a 13.

Write the missing number.

12 _ _ _ _ _ _ 14

Date: _____

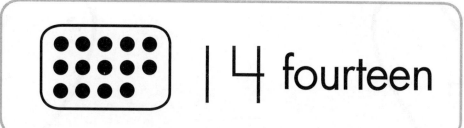

14 fourteen

Trace and write.

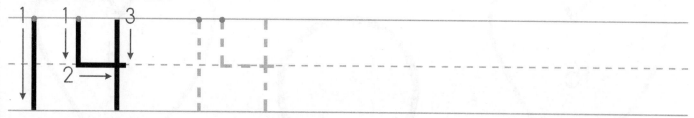

Color the set of 14 hearts.

Draw a line from the 14 to each heart with a 14.

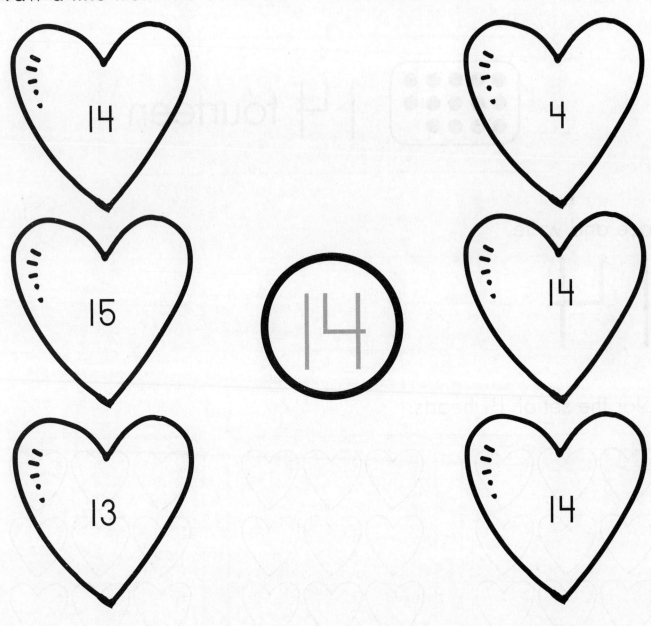

Write the missing number.

13 ----- 15

To parents   Give your child a long piece of yarn and show him how to make a simple knot. Ask him to knot the piece of yarn 14 times.

Trace and write.

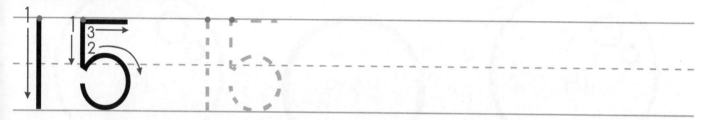

Color the bathtub with 15 bubbles.

Draw a line from the 15 to each bubble with a 15.

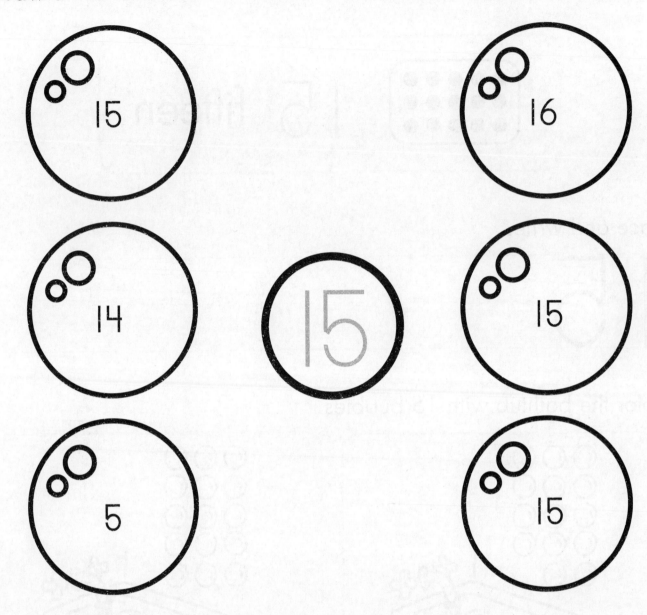

Write the missing number.

14 ___ 16

16 sixteen

Trace and write.

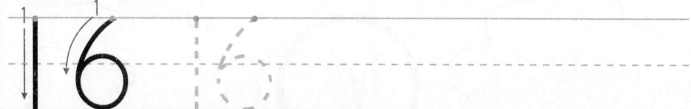

Color the cloud with 16 raindrops.

Draw a line from the 16 to each cloud with a 16.

Write the missing number.

15 —— 17

To parents Go on a walk with your child. Ask him to count the number of people wearing black shoes. Stop when he reaches **16**.

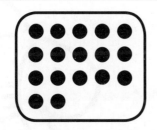 | 7 **seventeen**

Trace and write.

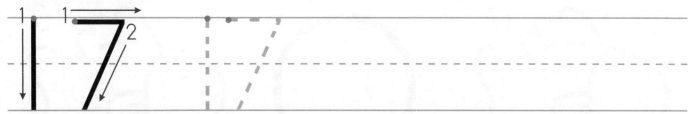

Color the set of 17 fish.

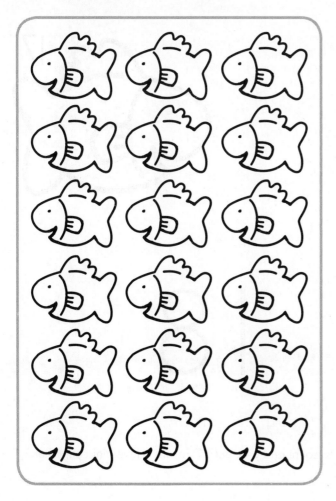 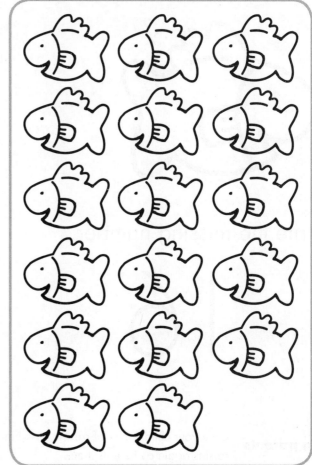

Draw a line from the 17 to each fish with a 17.

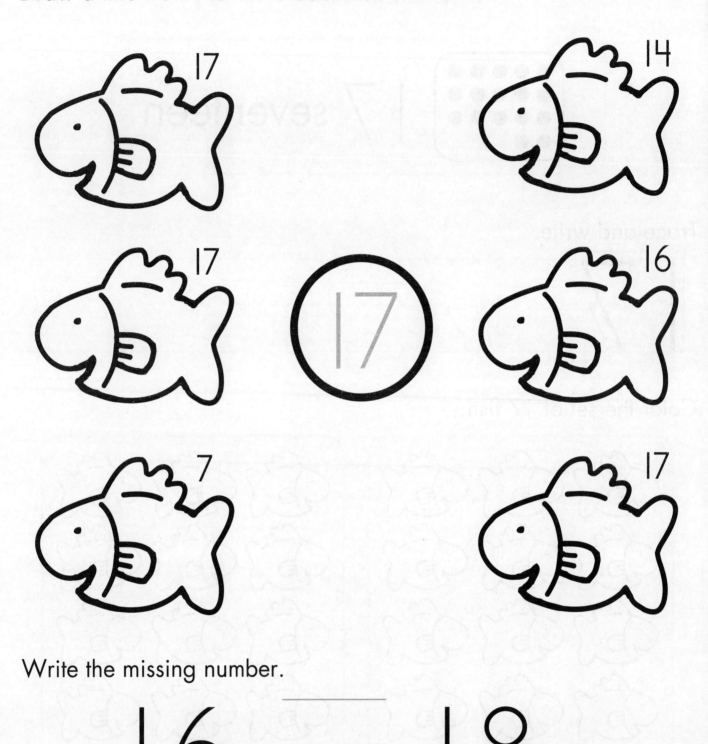

17

14

17

17

16

7

17

Write the missing number.

16 _____ 18

To parents — Give your child a box of raisins. Ask him to count and pick out **17** raisins. Have him count the raisins again as he eats them.

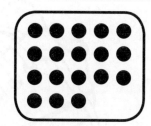 18 eighteen

Trace and write.

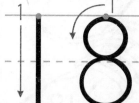

Color the set of 18 shoes.

 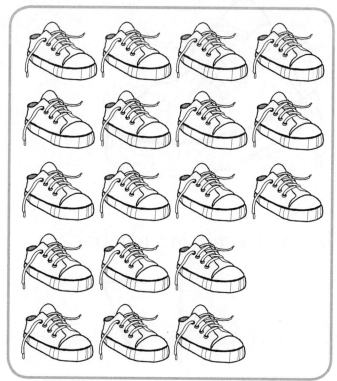

Draw a line from the 18 to each shoe with an 18.

Write the missing number.

17 ____ 19

To parents Go on a scavenger hunt around the house with your child and look for **18** items of the same color.

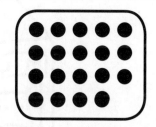 19 nineteen

Trace and write.

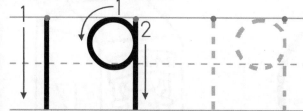

Color the set of 19 milk cartons.

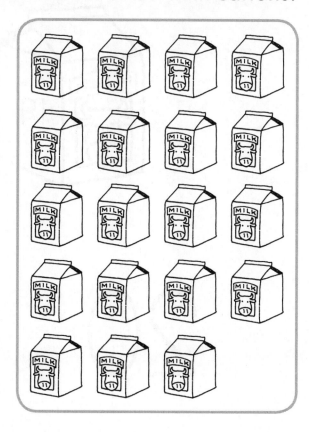

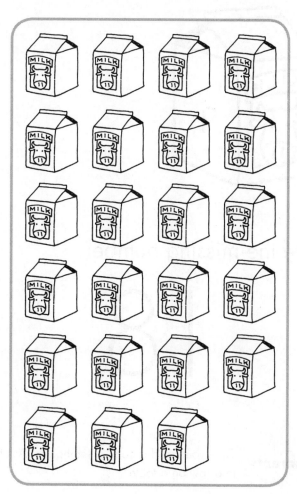

Draw a line from the 19 to each milk carton with a 19.

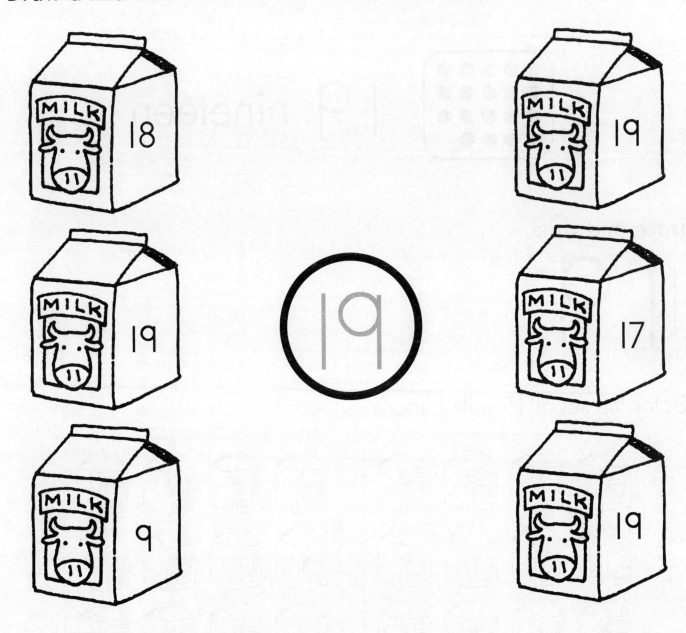

Write the missing number.

18 _ _ _ _ _ _ _ 20

To parents Give your child a pack of toothpicks. Ask him to count and take out **19** toothpicks, and place them in a row on a table.

 20   twenty

Trace and write.

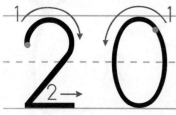

Color the set of 20 leaves.

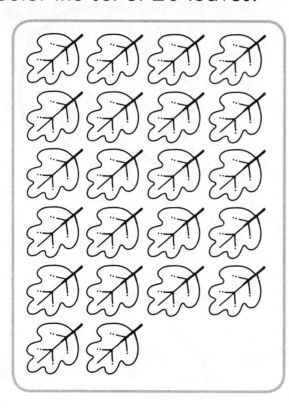

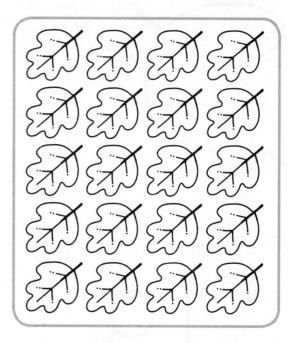

Draw a line from the 20 to each leaf with a 20.

Write the missing number.

19 _____ 21

To parents Draw a tree with ten red apples and another with ten green apples. Ask your child to count the number of apples on both the trees.

Review Numbers 11 to 20

Draw a path inside each number.

11	
	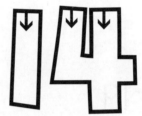

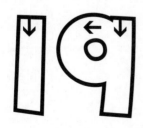	20

Connect the numbers 11 to 20 to make a picture. Color the picture.

11 START HERE

20

19

12

16

15

18

17

14

13

 21 twenty-one

Draw a path inside the numbers.

Color each circle with a 21.

21   22   21   24

  11   25   21   21

Draw a path for the car to follow. 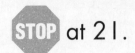 at 21.

1 2 3 4 5 6 7 8 9 10 11

12

13

20 19 18 17 16 15 14

21

22

23 24 25 26 27 28 29 30

Write the missing number.

20 ----- 22

To parents Have your child count and pick out **21** paper clips. Have him count the paper clips again as he links them together to make a chain.

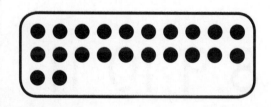

22 twenty-two

Draw a path inside the numbers.

Color each triangle with a 22.

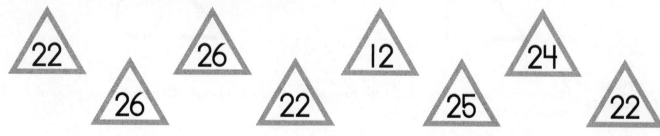

Draw a path for the car to follow. 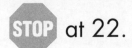 at 22.

1 2 3 4 5 6 7 8 9 10 11

12

13

20 19 18 17 16 15 14

21

22

23 24 25 26 27 28 29 30

Write the missing number.

21 _ _ _ _ _ 23

To parents Make a small photo frame using ice-cream sticks. Let your child paste **22** pieces of dry pasta all around the frame using craft glue.

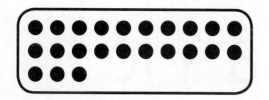

23 twenty-three

Draw a path inside the numbers.

Color each square with a 23.

| 23 | | 26 | | 25 | | 26 |
| 27 | | 23 | | 25 | | 13 |

Draw a path for the car to follow.  at 23.

1 2 3 4 5 6 7 8 9 10 11

12

13

20 19 18 17 16 15 14

21

22

23

24 25 26 27 28 29 30

Write the missing number.

22 _ _ _ _ _ 24

To parents Give your child a box of cherries. Ask him to count and pick out **23** cherries, and place them in a line on a clean table. Have him count the cherries again as you and your child share them.

 24 twenty-four

Draw a path inside the numbers.

Color each oval with a 24.

26   24   14   26

27   25   24   21

Draw a path for the car to follow.  at 24.

1 2 3 4 5 6 7 8 9 10 11

12

13

20 19 18 17 16 15 14

21

22

23 24 25 26 27 28 29 30

Write the missing number.

23 _____ 25

 25 twenty-five

Draw a path inside the numbers.

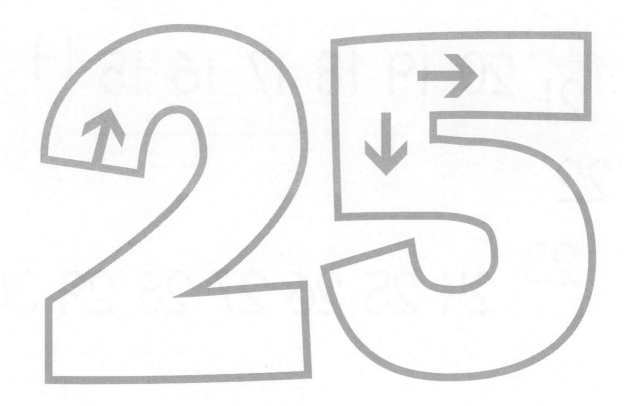

Color each diamond with a 25.

Draw a path for the car to follow.  at 25.

1 2 3 4 5 6 7 8 9 10 11

12

13

20 19 18 17 16 15 14

21

22

23

24 25 26 27 28 29 30

Write the missing number.

24 _ _ _ _ _ _ 26

To parents Stand at a safe place that overlooks a busy street. Have your child count the number of cars that go by. Stop when he reaches **25**.

Date: _____

26   twenty-six

Draw a path inside the numbers.

Color each rectangle with a 26.

| 25 | | 26 | | 15 | | 26 | |
| | 27 | | 25 | | 25 | | 22 |

59

Draw a path for the car to follow.  **STOP** at 26.

1 2 3 4 5 6 7 8 9 10 11

12

13

20 19 18 17 16 15 14

21

22

23 24 25 26 27 28 29 30

Write the missing number.

25 _____ 27

To parents Give your child a sponge cake and **26** pieces of cake decorations. Encourage him to decorate the cake.

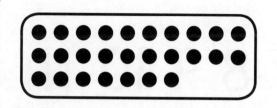 27 twenty-seven

Draw a path inside the numbers.

Color each arrow with a 27.

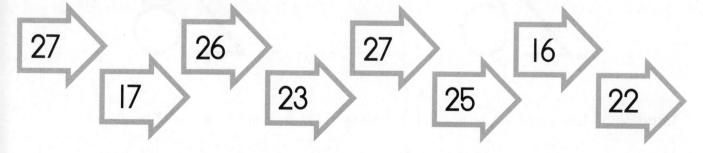

27   26   27   16

17   23   25   22

Draw a path for the car to follow.  at 27.

1 2 3 4 5 6 7 8 9 10 11

12

13

20 19 18 17 16 15 14

21

22

23 24 25 26 27 28 29 30

Write the missing number.

26 _____ 28

To parents   Take a small wooden bowl or tray. Have your child decorate it by pasting **27** beans on it.

Date: _____

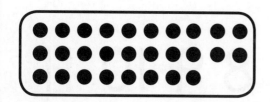 28 twenty-eight

Draw a path inside the numbers.

Color each heart with a 28.

28   26   27   29

18   19   28   22

Draw a path for the car to follow. 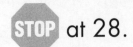 at 28.

1 2 3 4 5 6 7 8 9 10 11

12

13

20 19 18 17 16 15 14

21

22

23 24 25 26 27 28 29 30

Write the missing number.

27 ------- 29

To parents Give your child a sheet of A4-sized paper. Help him crease it **28** times to make a paper fan.

Date: _____

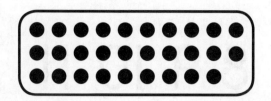

29 twenty-nine

Draw a path inside the numbers.

Color each star with a 29.

Draw a path for the car to follow. 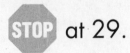 at 29.

1 2 3 4 5 6 7 8 9 10 11

12

13

21 20 19 18 17 16 15 14

22

23 24 25 26 27 28 29 30

Write the missing number.

28 _____ 30

To parents   Go on a scavenger hunt around the house, with your child, and look for **29** things whose names begin with any letter such as the letter B.

66

30   thirty

Draw a path inside the numbers.

Color each stop sign with a 30.

30   33   31   32

    20   10   30   30

Draw a path for the car to follow.  **STOP** at 30.

1 2 3 4 5 6 7 8 9 10 11

12

13

20 19 18 17 16 15 14

21

22

23 24 25 26 27 28 29 30

Write the missing number.

29 _____ 31

To parents Take a calendar and make a list of all the months that have **30** days. Go through the list with your child.

Review Numbers 21 to 30

Date: _____

Draw a path inside each number.

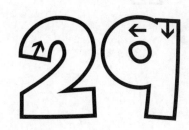

Connect the numbers 21 to 30 to make a picture. Color the picture.

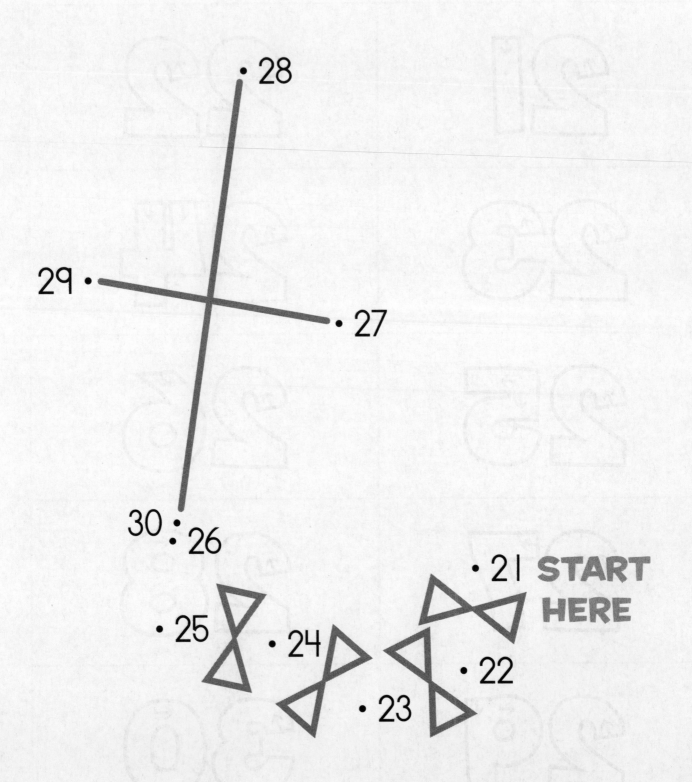

Review Numbers 1 to 30

Date: _____

Connect the numbers 1 to 30 to make a picture. Color the picture.

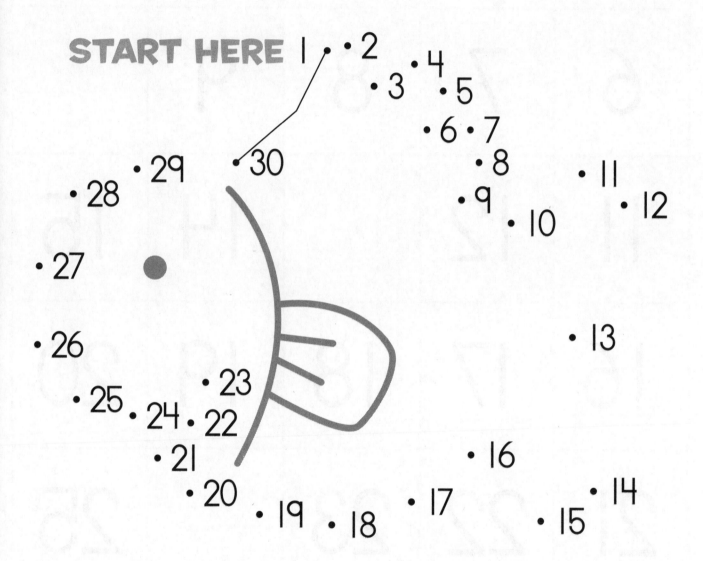

Fill in the missing numbers.

	2	3	4	
6	7	8	9	
11	12		14	15
16	17	18	19	20
21	22	23		25
	27	28	29	30

Draw a path inside the numbers.

Color each sun with a 100.

Draw 11 more suns to make 100.

Write the missing number.

Answer Key

Page 5

Page 10

2 3 4

Page 15

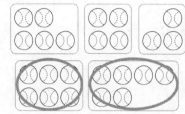

Page 6

I 2 3

Page 11

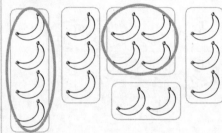

Page 16

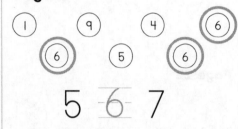

5 6 7

Page 7

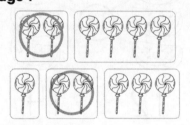

Page 12

3 4 5

Page 17

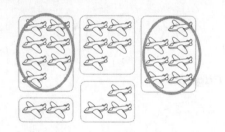

Page 8

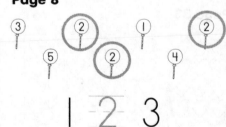

1 2 3

Page 13

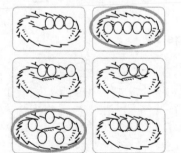

Page 18

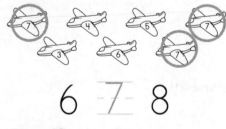

6 7 8

Page 14

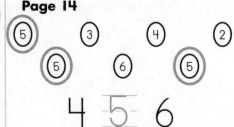

4 5 6

Page 9

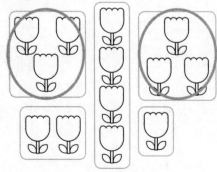

Page 19

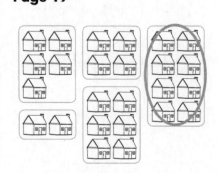

75

Page 20

7   8   9

Page 21

Page 22

8   9   10

Page 23

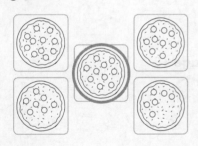

Page 24

9   10   11

Page 26

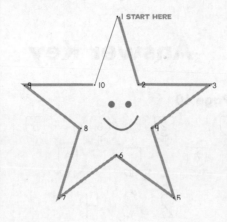

START HERE

Page 27

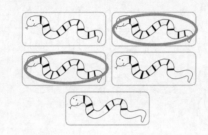

Page 28

10   11   12

Page 29

Page 30

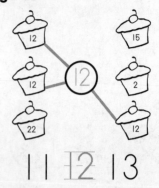

11   12   13

Page 31

Page 32

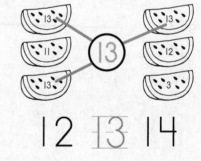

12   13   14

Page 33

Page 34

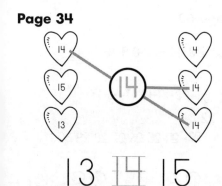

13 14 15

Page 35

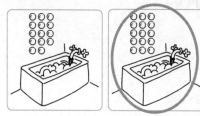

Page 36

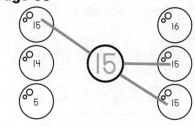

14 15 16

Page 37

Page 38

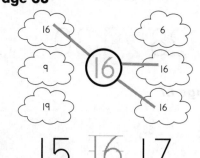

15 16 17

Page 39

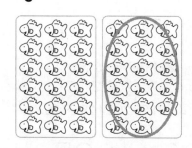

Page 40

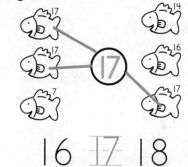

16 17 18

Page 41

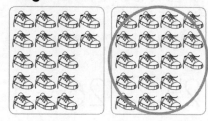

Page 42

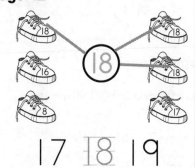

17 18 19

Page 43

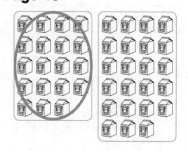

Page 44

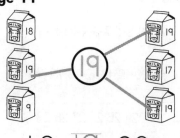

18 19 20

Page 45

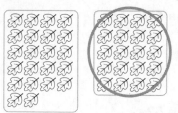

Page 46

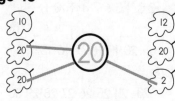

19 20 21

Page 48

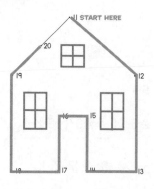

Page 49

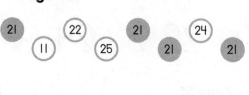

Page 50

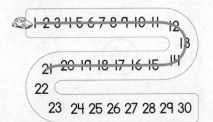

20 21 **22**

Page 51

Page 52

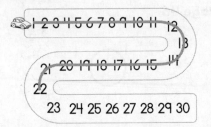

21 22 **23**

Page 53

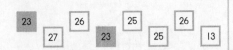

Page 54

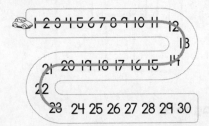

22 23 **24**

Page 55

Page 56

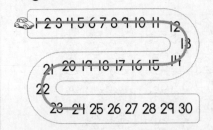

23 24 **25**

Page 57

Page 58

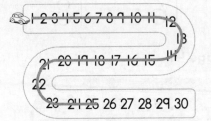

24 25 **26**

Page 59

Page 60

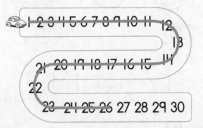

25 26 **27**

Page 61

Page 62

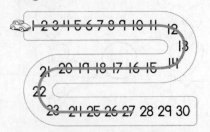

26 27 **28**

Page 63

Page 64

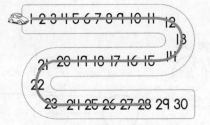

27 28 **29**

Page 65

Page 66

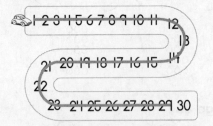

28 29 **30**

Page 67

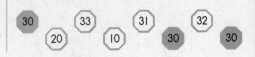

Page 68

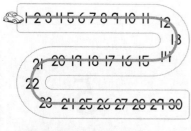

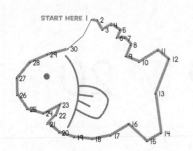

29 30 31

Page 70

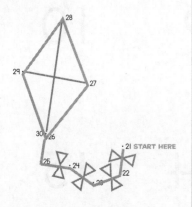

Page 71

START HERE

Page 72

1	2	3	4	5
6	7	8	9	10
11	12	13	14	15
16	17	18	19	20
21	22	23	24	25
26	27	28	29	30

Page 73

Page 74

99 100 101

Number Chart

1	2	3	4	5
6	7	8	9	10
11	12	13	14	15
16	17	18	19	20
21	22	23	24	25
26	27	28	29	30